P9-ARV-099

CHRIS HAYHURST

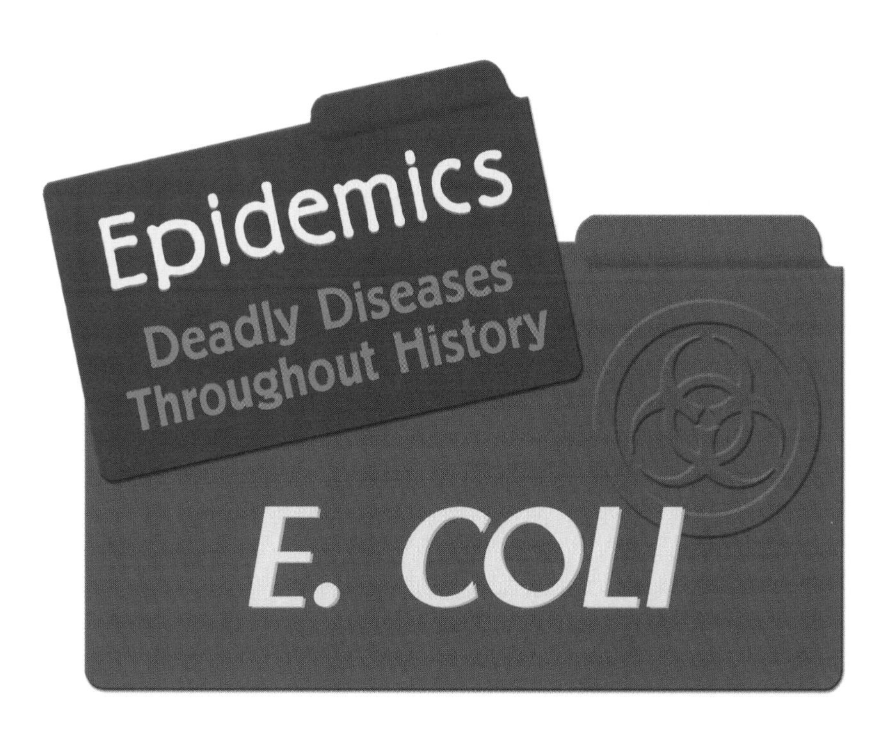

The Rosen Publishing Group, Inc.
New York

Published in 2004 by The Rosen Publishing Group, Inc.
29 East 21st Street, New York, NY 10010

First Edition

Library of Congress Cataloging-in-Publication Data

Hayhurst, Chris.
E. coli / by Chris Hayhurst.
p. cm. — (Epidemics)
Summary: Describes a dangerous new strain of *E. coli*, called 0157:H7, which appeared in 1982, how it is spread via contaminated meat, milk, vegetables, water, or person to person, the symptoms of the disease, and the methods used to handle outbreaks.
Includes bibliographical references and index.
ISBN 0-8239-4201-5 (lib. bdg.)
1. *Escherichia coli* 0157:H7—Juvenile literature. 2. *Escherichia coli* infections—Juvenile literature. [1. *E. coli* infections. 2. Diseases.] I. Title. II. Series.
RA644.E83 H39 2003
616'.01442—dc21

2003000751

Manufactured in the United States of America

Cover image: The cell wall of *Escherichia coli*, magnified 9000 times

CONTENTS

Workers at a Nebraska food plant process ground beef, one of the prime sources of E. coli *infection.*

INTRODUCTION

Alpine, Wyoming, is a very small town. Close to the state's western border, and home to fewer than five hundred year-round residents, it's the kind of place where just about everybody knows everybody else. Neighbors have each other over for outdoor barbecues. Children play together, go to church together, and attend school together. As is the case with most small towns, life is usually pretty normal in Alpine. Big news is rare.

But in the summer of 1998, things in Alpine were far from routine. First, near the end of June, a young woman and her five-year-old son became horribly ill. Then, around the same time, a motorcyclist just passing through also became sick. Within a few days, more than sixty people were complaining of the same symptoms: terrible stomach cramps and bloody diarrhea.

Local doctors quickly realized something was wrong. At first, they weren't sure what to think, but they had a hunch. Though they had never seen anything like this before, more and more people were coming down with the same sickness, and that was a sure sign that something bad was spreading around. They made a few calls and did a few tests, and before long the culprit was identified: *Escherichia coli* 0157:H7.

E. coli, as it's commonly called, is typically harmless. A microorganism found naturally in both human and animal intestines, it plays a vital role in digestion and helps the body absorb important vitamins from food. It also prevents the growth of dangerous bacterial species, acting much like a policeman for the intestines. There are hundreds of strains of *E. coli*. In day-to-day life, most of these strains are friendly. *E. coli* 0157:H7, however, is not.

E. coli 0157:H7 (the jumbled numbers and letters are used by scientists to describe the bacteria) is one of just a few *E. coli* strains capable of doing damage to the human body. *E. coli* 0157:H7 was first discovered in 1982 by scientists at the Centers for Disease Control and Prevention (CDC), a government agency that studies diseases and tries to prevent them. This particular strain has since become a worldwide cause of severe diarrhea and—in many cases—long-term illness and death.

To see why *E. coli* 0157:H7 can be so dangerous, it helps to first understand where the bacteria is found. It's also a good idea to know how the bacteria can find its way into the human digestive system and, of course, how doctors can diagnose and treat an *E. coli* infection. This book will go through all the things you need to know about *E. coli,* as well as a few things you just might find interesting.

While *E. coli* is certainly not a good thing to get, there's no need for people to panic in fear of contracting it. The truth is, *E. coli* 0157:H7 rarely kills, and it typically passes out of the body as quickly—if not as easily—as it enters. As the residents of Alpine, Wyoming, can testify, contracting the bacteria is bad, but it's certainly not a death sentence. In fact, not one of the victims of that outbreak died. Today, fortunately, life in Alpine is pretty much back to the way it was.

1

A HISTORY OF *E. COLI*

E. coli was first discovered in the 1800s by a man named Theodor Escherich. Escherich was a German bacteriologist, a scientist who studies bacteria. He found the bacteria living in the human colon, an important part of the digestive system. Through experimentation and close observation, Escherich eventually found that the bacteria could be blamed for ailments like diarrhea and other digestive problems. Today *Escherichia coli* is named in recognition of Dr. Escherich.

In the years following the discovery of *E. coli*, scientists learned a lot about it. One thing they found was that it was easy to make *E. coli* grow and reproduce very quickly. This was an important characteristic. It meant that laboratory workers could use *E. coli* to study many different things, and it wouldn't take a long time to get results. Today geneticists (scientists who study genes, which are the basic units of heredity) use *E. coli* all the time. They study *E. coli* under the microscope and learn all about its structure, how it reproduces, and how certain characteristics of the bacteria are passed on from generation to generation. These scientists can then take what they learn and apply it in ways to fight infections. Ultimately, perhaps, they'll discover a weakness in *E. coli* 0157:H7 that will allow them to develop specific treatments for those who suffer from it.

Dangerous Bacteria

More than twenty years ago, scientists first recognized they had a dangerous new strain of *E. coli*, a bacteria they had studied for years, on their hands. What they didn't realize was just how common this strain would ultimately become.

When *E. coli* 0157:H7 first appeared, it was in the United States. The first two outbreaks took place in Oregon and Michigan in February and June of 1982. The initial victims reacted in much the same way as people do today. They experienced terrible pains and cramps in their abdominal region—around the stomach. They had uncontrollable diarrhea, and it was often speckled with blood. They felt horribly ill, enough to think they might not survive. Some of the first victims went to their doctors thinking they would require immediate surgery and hospitalization. They thought they would die.

The doctors who first saw these people were not sure what to think. Their illness was unlike anything they had ever seen before. They ran tests for common diseases, but everything turned out negative. They could not get to the bottom of this strange sickness. As they struggled to come to a conclusion, more and more people fell seriously ill.

Soon the CDC stepped in. The CDC experts collected stool samples from the sick patients and began running a series of tests to see what they could find. They tested for *Salmonella*, a common bacteria that causes illness through the consumption of contaminated food. They tested for other known bacteria that caused similar symptoms, but they could not find anything that looked familiar. Then they made an interesting discovery. They found that a large number of the samples they were testing contained a new type of *E. coli*—one they had never seen before. They named this new strain *E. coli* 0157:H7. Then they switched the focus of their investigation.

E. COLI 0157:H7 STATS

- An estimated 73,000 cases of infection with *E. coli* 0157:H7 occur annually in the United States.
- *E. coli* causes an average of 61 deaths each year in the United States.
- Every year, 2,100 people are hospitalized in the United States because of *E. coli*.
- Infections from *E. coli* have been reported in more than thirty countries on six continents since 1982.

—Centers for Disease Control and Prevention

The scientists compared the stool samples of the ill people with those of healthy people—people who had not fallen ill. When they did so they found that the healthy people's stools had no traces of *E. coli* 0157:H7. This could mean only one thing: This strain of *E. coli* was not naturally found in the human body and had somehow invaded these people's intestines.

Eventually the CDC experts made another discovery. They found that all of the victims had eaten hamburgers at the same chain of fast-food restaurants. Normally such a discovery might not be so interesting. But when you're looking for clues as to why a certain population of people is sick with the same illness, it can be a key to solving the case.

BAD BACTERIA

E. coli is not the only organism worth avoiding when it comes to eating safely. There are many other common bugs and worms capable of making people sick. Here's a partial list of the worst:

- *Salmonella*: Found in eggs, chicken, and fresh produce, it causes severe diarrhea, much like *E. coli*. *Salmonella* can cause bone infections in people suffering from sickle-cell anemia, a disease of the blood.
- *Trichinella spiralis*: This worm infests undercooked or raw pork. Typical symptoms are similar to the flu—victims might get a fever or just feel run-down. In extreme cases, *Trichinella spiralis* can eventually cause heart and breathing problems.
- *Vibrio*: Found in raw shellfish, this leads to flu-like symptoms and can cause death.

Suspicious, the experts tested samples of frozen ground beef from these restaurants and found that they, too, contained *E. coli* 0157:H7. At this point it didn't take much to realize what had happened. The sick people had eaten hamburger meat tainted with *E. coli*, and now that *E. coli* had made a home in their bodies. It was destroying their digestive systems and was now present in their stools. Somehow *E. coli* 0157:H7 was causing these people—forty-seven in all—to be sick.

Round Two

It was just a few months after this nasty introduction to *E. coli* 0157:H7 that health experts had their hands full once again. But this time the outbreak was in a different country. The people who were sick were in Ontario, Canada. The victims were residents of a home for the elderly. They all lived together and were taken care of by staff members. The staff would help them in their day-to-day lives, with things like bathing, dressing, and cooking. The residents went about their daily lives just as they had for years. But then, in November 1982, dozens of people became sick. In all, 31 of the 353 people at the home fell ill. Their symptoms were telltale, especially if you were an expert on *E. coli* 0157:H7: painful stomach cramps and bloody diarrhea.

Scientists known as epidemiologists were called to the scene. Epidemiologists are experts at studying and tracking diseases in specific populations—like the unique population of people at the rest home. The epidemiologists ran their standard tests. Before long they found that *E. coli* 0157:H7 was present in more than half of the victims' stools. They then traced the contamination to a meal that had just been served to the home's residents. That meal, not surprisingly, included hamburgers. But they

SYMPTOMS OF *E. COLI*

Bloody diarrhea and abdominal cramps are the most telling symptoms of an *E. coli* infection, but not everyone experiences such discomfort. Sometimes the diarrhea has no blood at all, and sometimes no symptoms whatsoever are felt. Fever is rare in *E. coli* victims, but when it does occur it's usually not too high. Typically the bug is out of the system within five to ten days.

also found something mysterious and especially frightening. Some of the victims had not eaten any of the food from that meal. Somehow these people had wound up with *E. coli* in their systems without putting one bite of contaminated hamburger meat into their mouths.

Investigating the case in more detail, the scientists eventually determined that in a few instances the bacteria had been spread from person to person. The only way this could have happened, they concluded, was if hygiene was not adequate among the residents. That is, if people were sick with *E. coli* and had diarrhea and then they failed to clean up sufficiently, it was possible for other people to make contact with the diarrhea and also become sick. Because the *E. coli* microorganism is so small, even the slightest trace of it on one person's hand could easily make its way into someone else's

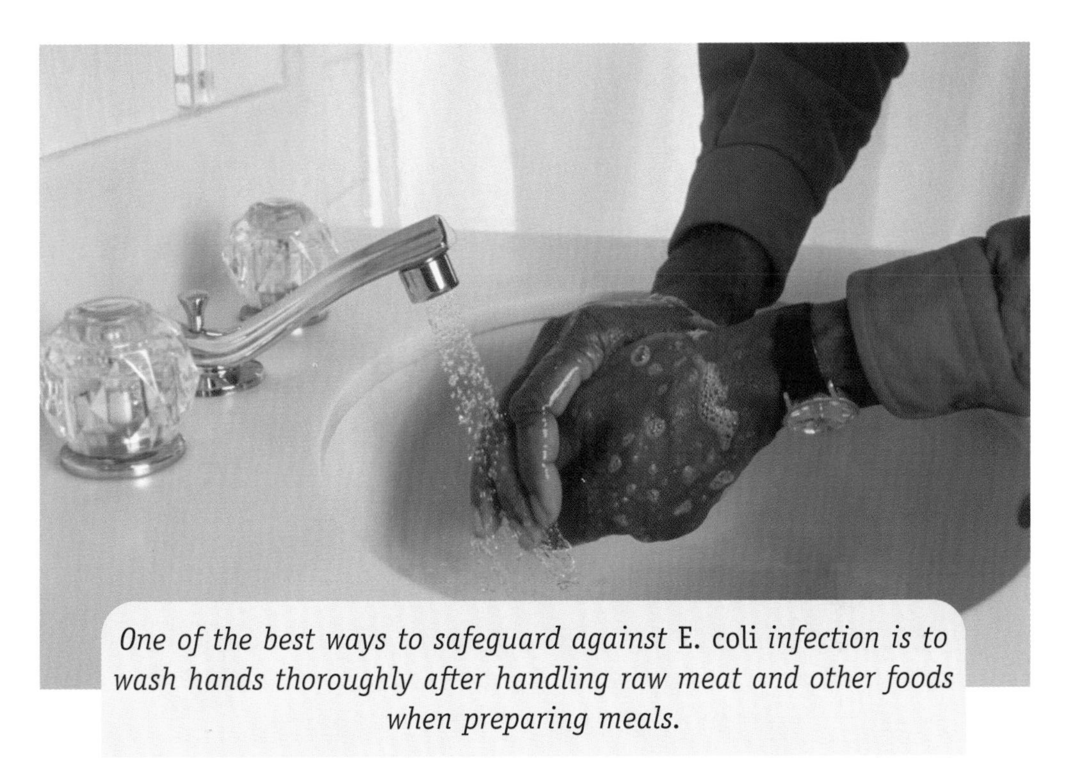

One of the best ways to safeguard against E. coli *infection is to wash hands thoroughly after handling raw meat and other foods when preparing meals.*

stomach. This strange new bacteria called *E. coli* 0157:H7 had suddenly revealed itself to be even more dangerous.

The Years Since

Since those first few outbreaks of *E. coli*, the bacteria has turned up quite regularly throughout the United States, Canada, and the rest of the world. Most cases have been isolated—limited to just one or two unfortunate people who happened to eat contaminated food. But major outbreaks have occurred, and when they do the damage can be enormous.

Though E. coli *and other infectious agents can be tracked to an extent, the massive distribution of meat throughout the world can also make widespread outbreaks hard to trace.*

The largest outbreak in the United States took place between November 1992 and February 1993. Four states were involved—first Washington, then Idaho, California, and Nevada—and ultimately more than five hundred people came down with *E. coli* infections. Four people eventually died.

Scientists investigating this huge outbreak determined that everything began at a single fast-food restaurant chain that served hamburgers. They traced the food back to its source and found that all the hamburger meat came from the same five slaughterhouses in the United States and Canada. The contaminated meat had been delivered to the restaurants. Once at the restaurants, it was not cooked well enough before it was served to the customers. Raw meat was to blame. Health experts quickly issued a recall of all meat supplied by those five slaughterhouses and stopped the outbreak in its tracks.

But not all *E. coli* outbreaks are caused by meat. The year after those five hundred people became sick from eating hamburgers, an outbreak took place overseas in Scotland. This time cows were involved, but milk—not beef—was to blame. Most of the people who became sick—and there were seventy in all—drank milk from the same local dairy. Amazingly, the milk had been pasteurized. Pasteurization is a heating process designed to kill any dangerous microorganisms present in the milk, so investigators were puzzled by how the contamination could have occurred. No one had ever become sick with *E. coli* from drinking pasteurized milk before. So why would people be sick now?

Part of the answer was found when the scientists investigating the case discovered traces of *E. coli* in a pipe that carried milk from the pasteurization building to the bottling machine. They also found small amounts of *E. coli* on the bottling equipment itself and in a tank used to hold the milk. Finally, they found *E. coli* in some of the cow pastures, indicating that the whole ordeal had begun with the cows. The investigators never did make a final determination of exactly how the milk became contaminated, but they have several theories. One theory is that the pasteurization process was incomplete. Perhaps the microorganisms were able to survive the pasteurization if, for instance, the milk was not heated to a high enough temperature. Another theory is that the milk somehow became contaminated after it was pasteurized—somewhere between the pasteurizing building and the bottling machine. Either way, the victims in Scotland were not happy.

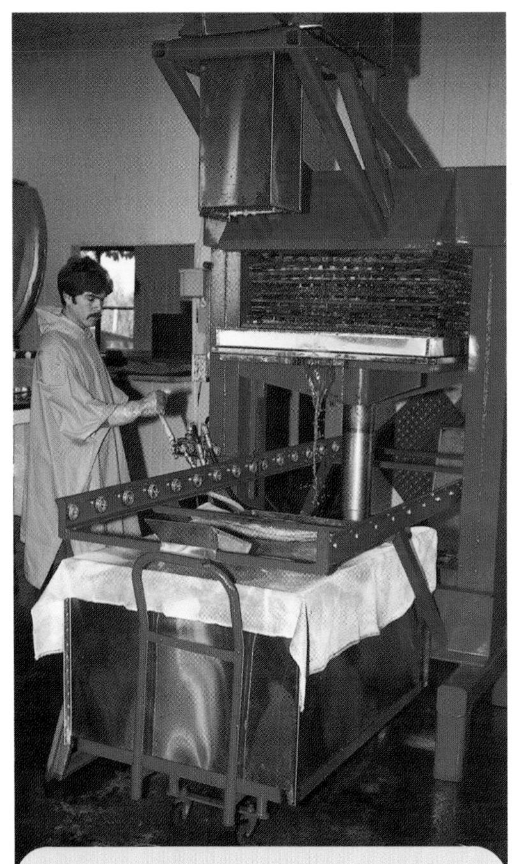

Even apples, like the ones being pressed into cider on this machine, can spread E. coli *to human beings.*

Another interesting outbreak took place in Massachusetts in 1991. This time the culprit was contaminated apple juice. Cows lived on the same farm where the apples were grown. The apples became contaminated before they were crushed into juice, most likely when they fell on the ground and made contact with cow feces. When the apples were crushed, the *E. coli* spread from the surface of the apples and was mixed into every last drop of the juice. People then contracted *E. coli* when they drank the juice. More than twenty people became sick.

2

LESS FAMOUS OUTBREAKS

Cases of *E. coli* involve just a few people and wind up receiving very little press. After all, one or two sick people with stomachaches and diarrhea is hardly reason for newspapers to publish front-page reports. They'd rather focus on big news events, like the major outbreaks where hundreds become ill. Still, when *E. coli* does make it into the human food chain, those affected certainly notice. The following are just a few of the *E. coli* outbreaks reported to the U.S. Food and Drug Administration from years past.

Seattle, Washington, November 16–December 21, 1994

A total of twenty cases of *E. coli* are confirmed through laboratory testing for the bacteria.

Epidemiologists investigating the case determine the infections occurred when the victims ate contaminated dry-cured salami. Later, three more people fall ill in northern California.

Rockford, Illinois, July 5, 1995

Five children are hospitalized with *E. coli* infections. Officials at the Winnebago County Health Center look for a common food source but come up empty. Further investigation reveals all of the children had gone swimming at the same state beach. The investigators decide to close the beach for fear that the children had contracted the *E. coli* by accidentally drinking contaminated lake water.

British Columbia, California, and Washington, October 1996

An outbreak of *E. coli* O157:H7 is reported in Washington State on October 30, 1996. Washington State Department of Health officials determine the victims all drank a particular brand of unpasteurized apple juice. Eventually a total of forty-five cases are reported not only in Washington but also in nearby California, Colorado, and British Columbia, Canada. Most of those stricken with the *E. coli* are very young (the average age is five years old). One of the patients has not drunk any

1885
Theodor Escherich, a bacteriologist in Germany, discovers a strange bacteria living in human intestines; the bacteria is later named *Escherichia coli*, or *E. coli*.

1972
The U.S. Congress passes the Clean Water Act, requiring wastewater treatment plants to prevent polluted water from entering the supply of drinking water.

1982
Two outbreaks of intestinal illness occur in Oregon and Michigan in February and June; scientists at the CDC determine the cause is *E. coli* 0157:H7.

1982–Present
E. coli outbreaks have been reported in more than thirty countries on six continents.

1991
An *E. coli* outbreak due to contaminated apple juice in Massachusetts sickens twenty people.

1992–1993
The largest outbreak in the United States takes place between November 1992 and February 1993 in Washington, Idaho, California, and Nevada; fifty-five people fall ill and four die.

1994
An *E. coli* outbreak in Scotland is traced back to contaminated milk. Seventy people fall ill.

1995–Present
E. coli infections continue to pop up throughout the United States and Canada.

apple juice at all, but contracted *E. coli* through secondary transmission from another victim.

The juice company, once informed that it is responsible for the outbreak, voluntarily issues a recall of all its apple juice products, preventing further spread of the bacteria.

West-Central Wisconsin, June 8–June 12, 1998

Eight cases are confirmed by laboratory tests by the Division of Public Health, Wisconsin Department of Health and Family Services. The *E. coli* source is fresh cheese curds from a local dairy.

Tarrant County, Texas, June 9–11, 1999

Teenagers attending a summer cheerleading camp fall ill. They experience nausea, as well as other symptoms, including vomiting, abdominal cramps, and diarrhea. A few of the cheerleaders have blood in their diarrhea. Stools are analyzed by local authorities. Nothing is found. The stools are sent to the Texas Department of Health and the CDC. *E. coli* O157:H7 is found in two of the samples.

Albany, New York, September 3, 1999

Ten children are hospitalized after attending a local county fair. The primary symptom is bloody diarrhea.

CLOSE CALL

In September 2002, a meatpacking plant in Pennsylvania recalled nearly 204,000 pounds of fresh ground beef. The company was worried the beef might be contaminated with *E. coli*. Unfortunately for the plant, the meat had already been sent out to eleven states. Luckily there were no reported illnesses.

Investigators eventually discover *E. coli* 0157:H7 in their stool.

E. coli in the Water

Many of the minor outbreaks of *E. coli* were not a result of people eating contaminated meat. What are other ways that people can be exposed? Studies have shown that *E. coli* 0157:H7 is able to survive for several months in fresh water. It often enters the water by runoff from farmland that has been fertilized with sewage sludge. It can also get into the water from septic tanks or leaky sewers. Once the *E. coli* is in surface water—such as streams, rivers, and lakes—it can then make its way to shorelines along the coast. The presence of *E. coli* in these places is not good news for water lovers. Many people like to use lakes for swimming, sailing, canoeing, or windsurfing. It's easy to swallow water while participating in these activities.

Outdoor pools can also become contaminated with *E. coli*. One way this could happen is when people

While *E. coli* most often passes into the human bloodstream through contaminated meat, it can also do so through drinking water. Waterborne outbreaks of *E. coli* are not common, but they do occur.

Most tap water is perfectly safe to drink. Years ago, this was not the case. But in 1972 the United States Congress passed a law called the Clean Water Act. The Clean Water Act requires cities to use wastewater treatment plants to prevent polluted water from making its way into its citizens' houses. Today specific rules are designed to prevent water contamination with bacteria like *E. coli*, which could potentially be present if the water contained fecal matter. The Environmental Protection Agency (EPA) requires cities and towns to filter and disinfect water before it is distributed.

It can be fun to find out exactly what's in the water that pours from your tap. Laboratories throughout the country offer water-testing services where professionals will examine your water and come up with a list of the various things they find. For a list of state-certified laboratories, as well as basic drinking-water information, call the EPA's Safe Drinking Water Hotline at (800) 426-4791. Or, you can go to their Web site at www.epa.gov/safewater. Another way to get the scoop on your water is by talking to the folks at your local, county, or state health departments. In some areas these health departments offer free water-testing services. You can look them up in the government pages of your phone book.

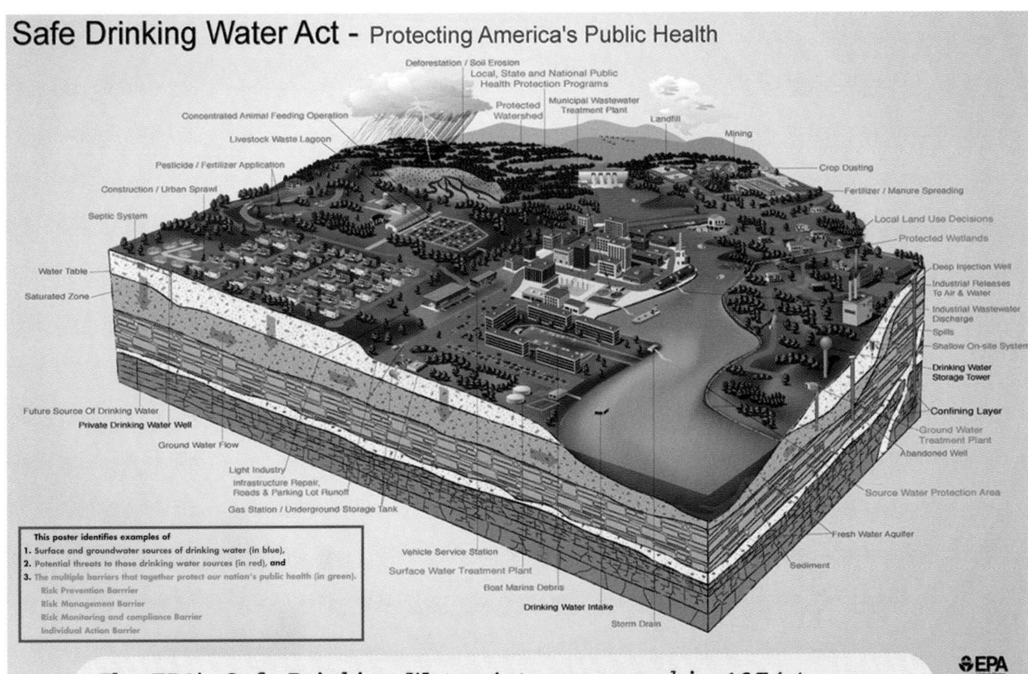

The EPA's Safe Drinking Water Act was passed in 1974 to ensure the quality of drinking water in the United States. This diagram details the everyday factors that affect your drinking water.

walk into the pool with dirty feet. Or an animal might wind up in the pool accidentally, and thus contaminate the water. However, most pools are chlorinated. Chlorine is a chemical that can kill microorganisms, including *E. coli*.

The waters of many popular beaches are regularly checked for contamination with *E. coli* and other organisms. If anything dangerous is detected, the public health authorities issue an alert. That alert might take the form of a sign at the entrance to the beach. So if you see a sign that says, "Danger, No Swimming," think twice about ignoring it. It might be there for a reason.

Beyond Beef

Few people realize that the *E. coli* bacteria can contaminate more than just beef. Thanks to spectacular reports of *E. coli* outbreaks due to beef contamination and the subsequent recalls of thousands of pounds of meat from grocery stores across the nation, it's easy to think that beef is the only danger when it comes to *E. coli*. However, the truth is that *E. coli* can be found on vegetables as well. In fact, while the overwhelming majority of outbreaks are due to beef contamination, experts believe there have been at least seventeen *E. coli* outbreaks in the United States since 1990 related entirely to contaminated produce.

One of the most recent produce-related outbreaks took place during the summer of 2002 in Spokane, Washington. That July at least thirty-four people attending a camp for cheerleaders at Eastern Washington University became sick when they ate contaminated romaine lettuce during dinner. The lettuce was part of an otherwise tasty Caesar salad. Experts believe the *E. coli* bacteria somehow hitched a ride from the farm to the kitchen and then made it to the campers' plates when the lettuce was not washed thoroughly. By the time the outbreak was in full swing, at least one of the campers became so ill

Meat is not the only way E. coli *can be transmitted. Produce can also be contaminated, which is why it is important to make sure everything, including lettuce, is cleaned properly.*

that her kidneys were affected and she was forced to go on dialysis, a medical procedure where a person's blood is purified by machines.

Today it is somewhat rare for a full-fledged outbreak to occur. Epidemiologists have learned a lot by studying past outbreaks, and they have used what they've learned to help prevent new ones. People do come down with *E. coli*, but usually the cases are limited to a few people. And if scientists and doctors have anything to do with it, an outbreak the size of the one in the United States back in 1992 and 1993 will never happen again.

3

THE SCIENCE OF THE SICKNESS

To see why *E. coli* 0157:H7 can be so dangerous, it helps to first understand how the bacteria can find its way into the human digestive system. Eating contaminated meat, vegetables, or fruit is the most common way, but other routes are also possible. *E. coli* is unique in that infection by as few as ten microorganisms can lead to symptoms. Most other bacteria must be present in large numbers before their effects become noticeable. This is one of the reasons *E. coli* is so dangerous—it doesn't take a lot to cause trouble!

From Cows to Burgers . . .

Most experts agree that *E. coli* 0157:H7's main route of entry is through the consumption of contaminated meat. *E. coli* is found on many cattle farms, where it lives in the stomachs of healthy cows. When the cows are slaughtered, the process can be messy. Sometimes the bacteria makes contact with the meat. Later, when that meat is chopped and blended into ground beef, the *E. coli* is ground into the mixture, too. If someone then eats that ground beef—say, as a grilled hamburger—the bacteria can make its way from the meat and into that person's stomach.

. . . and Milk

Because cows can carry *E. coli*, the bacteria can be spread through milk, too. Most milk is sterilized through pasteurization. Pasteurization destroys all of the dangerous bacteria that might be present in the milk. Some milk, however, is not pasteurized. Unpasteurized milk is referred to as "raw milk." Some people like raw milk because they believe it tastes better and think it's better for them in its natural state—straight from the cow. But if there is *E. coli* on the cows' udders or on the milking equipment, it's possible for the bacteria to enter the milk. If someone drinks the milk, he or she is at risk of contracting *E. coli*.

Fruits and Vegetables

Another way *E. coli* is spread is through the consumption of unwashed fruit, vegetables, and other raw foods. If the food has been mishandled somehow—by someone who failed to wash his or her hands after going to the bathroom, for instance—or if it came in contact with contaminated meat, the *E. coli* can live on the food's surface. If it does and it is eaten by someone, that person can contract *E. coli* and become sick.

Other Methods of Contraction

While most cases of *E. coli* occur through the consumption of contaminated food, the bacteria can also be spread in other ways. Children have contracted *E. coli* by putting contaminated dirt in their mouths. Swimmers have become infected when they accidentally swallowed lake water contaminated with fecal matter. Rest homes have reported outbreaks due to inadequate post-bathroom hygiene among their elderly residents. And drinking water has been implicated in a number of cases, including the one that paralyzed Alpine, Wyoming, in 1998. Once the bacteria is in the water, it can make its way to people's taps if water-treatment facilities are not working correctly.

Cross-contamination can cause the spread of E. coli *in places where food is prepared for public consumption, such as delicatessens and butcher shops.*

People can also become sick from *E. coli* when the food products they eat have been "cross-contaminated" by other food products. Cross-contamination takes place when a food that has *E. coli* on it makes contact with a food that does not. For instance, cooked sliced meats and cooked meat products for sandwiches and salads have become contaminated when they made contact with uncooked meats that carried the bacteria. The contact usually takes place accidentally in places like butcher shops and delicatessens.

Finally, person-to-person outbreaks have occurred, too. Those most at risk are very young children and

E. COLI AND KIDS

Toddlers can carry *E. coli* in their feces for up to two weeks after they stop showing symptoms of the infection. Older kids, on the other hand, usually start shedding the organism around the time their symptoms go away. For this reason it's a good idea to be extremely careful around very young children who have recently been infected by *E. coli*, as traces of the bacteria can show up on toys, blankets, clothing, or anything else that has been touched—especially if the child's hands are not clean.

people at institutions where personal hygiene is not good, like in psychiatric wards, homes for the elderly, and children's day-care facilities. In day care, infection is usually introduced by a sick child. Because young children may not wash their hands after going to the bathroom or may otherwise be messy, an outbreak can occur. All it would take is for one child to get *E. coli* from a food product. He or she could then spread the infection to other children. Experts recommend that young children with diarrhea should be kept home from day care until they've recovered.

In the Body

Once it's in the human body, *E. coli* 0157:H7 takes on a mind of its own. It enters the intestines and destroys the intestinal blood vessels. This causes the victims to bleed when they go to the bathroom. Bloody diarrhea is one of the key signs doctors look for when they diagnose a person with an *E. coli* infection.

In more severe cases, the *E. coli* toxin then travels into the victim's bloodstream. Once in the bloodstream, it can spread throughout the body and cause more destruction. Various blood vessels break apart. As the blood vessels break, blood clots form. Blood clots clog up blood vessels. If they clog the wrong places, they can be very dangerous.

One of the main problems with *E. coli* is it often leads to damage of organs, such as the heart, the lungs, and the kidneys. In fact, it is kidney problems—specifically a disorder known as hemolytic uremic syndrome, or HUS—that most often lead to death in *E. coli* victims. Kidneys are important for the excretion of unwanted waste products from the human body. HUS is most common in elderly people and in children under the age of five. HUS leads to destruction of cells of the body. This condition is linked to kidney problems. If the kidneys stop working, the body can't excrete its waste products.

Studies show that almost one-third of those with HUS continue having kidney and other problems many years after their initial infection with *E. coli*.

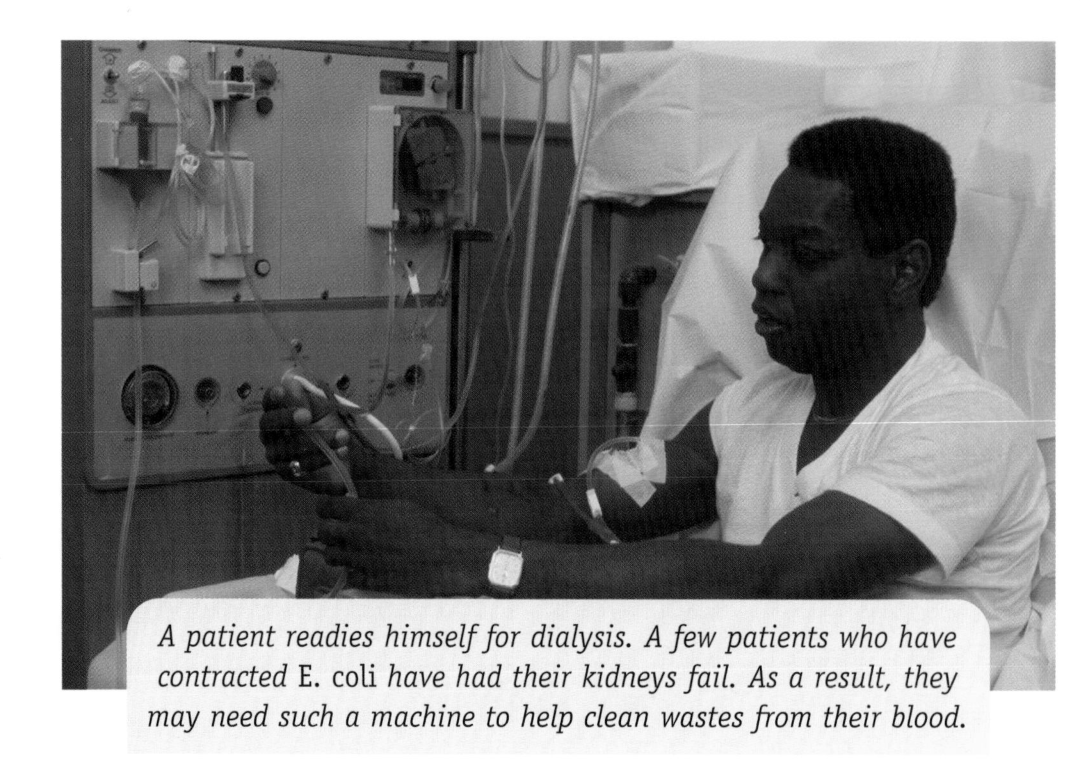

A patient readies himself for dialysis. A few patients who have contracted E. coli *have had their kidneys fail. As a result, they may need such a machine to help clean wastes from their blood.*

For some, their kidneys fail to function altogether. This is known as renal failure. When renal failure occurs, dialysis becomes necessary. Dialysis machines clean the wastes out of the victim's blood and help bring the concentration of important body chemicals to normal levels. Other HUS victims experience medical conditions such as high blood pressure, seizures, blindness, and paralysis. Despite all the drawbacks of developing HUS, the syndrome doesn't necessarily lead to death. In fact, less than 5 percent of those who develop HUS and receive proper medical care die from the condition.

Tracking the Spread of *E. coli*

Today's battle against *E. coli* is incredibly sophisticated. Epidemiologists try to learn everything they can about a disease—where it most commonly occurs, who it affects, how to control it, and if they're lucky, how to destroy it. They do everything possible to stop diseases in their tracks and are often the first scientists on the scene when an outbreak occurs. In the United States, the epidemiologists leading the fight against *E. coli* work at the CDC in Atlanta, Georgia. These scientists study not only *E. coli* but also other serious agents like HIV (the virus that can lead to AIDS), *Salmonella*, and diseases such as cholera. Epidemiologists from other health organizations—in the United States and throughout the world—also study these diseases.

One of the main jobs of today's epidemiologist is to track the pathogen—the cause of a disease—back to its source. In the case of *E. coli*, doing this efficiently can be extremely difficult. Interviews must be conducted with victims, their families, and their doctors. Questions are asked about what they ate and when. What did they drink? Do they remember exactly where they ate, what they ate, and the way the food was prepared, cooked, and served? The answers to these questions can supply clues for the

DID YOU KNOW?

Bacterial cells like *E. coli* reproduce through a process called binary fission. In binary fission, a parent cell essentially splits in half. When the process is over, there are two identical cells where before there was one.

Binary fission is an extremely efficient way of reproducing. The two bacteria resulting from the first split can each reproduce to make a total of four bacteria. Those four can do the same, making eight. The process can then continue again and again, resulting in billions if not trillions of offspring from just a single cell.

Scientific studies have determined that *E. coli* can divide about every twenty minutes. This means that a single *E. coli* cell grown in the lab can produce a colony of between 10 million and 100 million bacteria in just twelve hours. In the human body, *E. coli* divides and replicates fast enough to create 20 billion new bacteria every day. That's just enough to replace the number lost to the environment through bodily wastes.

epidemiologists to use to help solve the case. By asking questions of all the victims of an outbreak, it's possible to quickly and efficiently find the source of the outbreak.

Another thing the epidemiologists do is collect and analyze stool and blood samples from the victims. They collect these samples and then bring them back to laboratories to be studied. At the laboratory the samples are analyzed for their chemical makeup, and they are examined under a microscope for telltale signs of the *E. coli* bacteria.

Finally, once everything is tested and all the interviews are done, the investigators often turn to their computers for help. Powerful programs can analyze all the numbers and information and come up with reasonable guesses as to the cause of the outbreak and its source. Of course, computers can't do it all, so these statistical analyses are used in addition to the work the investigators do by hand.

Once the outbreak is tracked to its source, the next step is containment. Health agencies will put out an alert to beef suppliers, for example, to let them know that a particular source of meat appears to be at the heart of the problem. The job of the epidemiologists at this point is to destroy the bacteria before it can claim more victims.

Each time epidemiologists confront an *E. coli* outbreak—or, for that matter, an outbreak of any disease—they use what they learn along the way to hopefully reduce the risk that similar outbreaks will occur in the future. They study exactly how the bacteria made its way into the human food chain and then try to set up new barriers to prevent it from taking that same route again. They also work with local public health organizations to see if there are any weaknesses that might be improved upon. The goal is to make sure the same thing does not happen again—especially in the same place.

4

CURRENT RESEARCH AND CURES

While *E. coli* 0157:H7 has, over the years, become a major concern of health professionals, consumers need to develop just a few simple habits to significantly reduce their chances of falling ill. One of the main things doctors and scientists recommend people do to prevent the spread of *E. coli* is to be careful while cooking.

Unfortunately, you can't tell if meat contains *E. coli* just by looking at it. It usually looks and smells normal. The only way to be sure meat is safe to eat is by complete cooking. Thorough cooking destroys the disease-causing *E. coli* bacteria. The CDC recommends using a thermometer to make sure all the meat—even the inside—is completely cooked. Only once all parts of a beef patty reach at least 160°F can you be sure the meat is completely cooked. If a thermometer is not available, experts recommend making sure there are no pink spots in the meat. Pink spots indicate the meat is still raw and has not been cooked enough.

You don't necessarily have to eat beef to get *E. coli*. Often the bacteria spreads when raw meat is chopped on a cutting board and then the board is not cleaned before vegetables or other foods are placed on it. Hands, countertops, and all utensils should be washed with hot, soapy water if they come in contact with raw meat. The main point is to not let any trace of raw meat make it into your mouth. So if anything comes in contact with raw meat, it must first be cleaned before it can be eaten or used in other food preparation.

It is very important to make sure meat is cooked properly in order to avoid getting sick from eating it.

Experts also recommend drinking only pasteurized milk and juices. The pasteurizing process kills dangerous

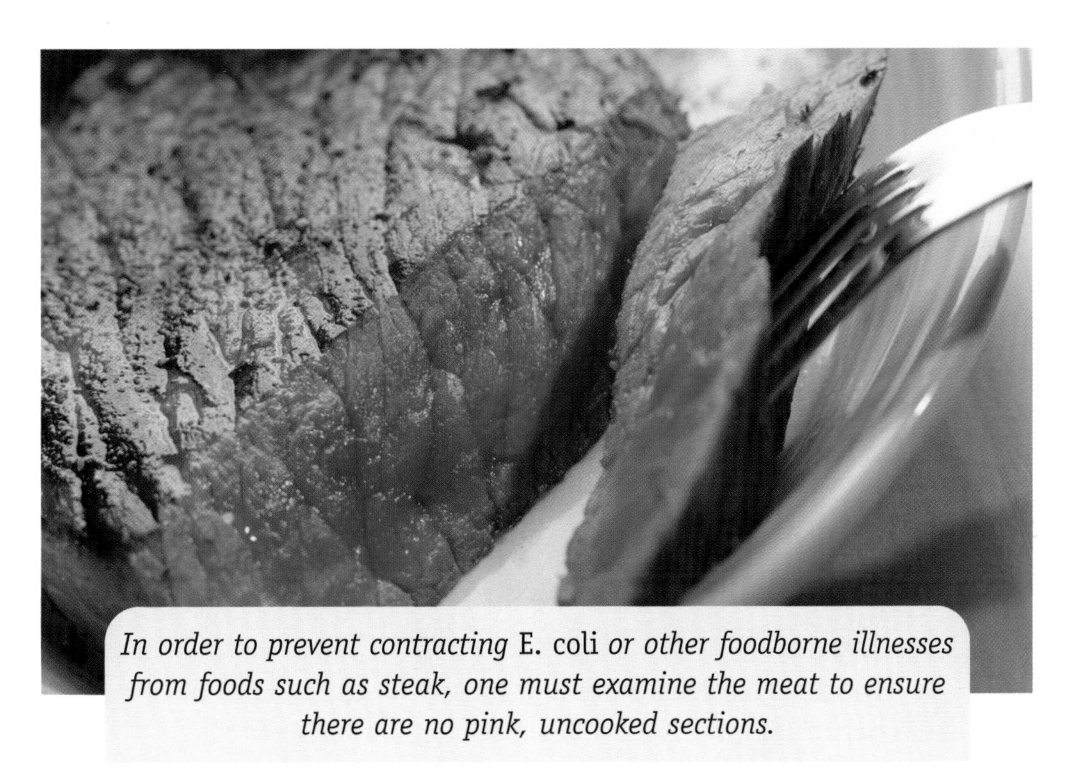

In order to prevent contracting E. coli *or other foodborne illnesses from foods such as steak, one must examine the meat to ensure there are no pink, uncooked sections.*

bacteria like *E. coli*. Most milk and juices sold in grocery stores are pasteurized.

Finally, fruits and vegetables should also be washed thoroughly before they are eaten, as they may become contaminated with *E. coli* through the soil or water or by coming in contact with animal feces or meat during the shipping process.

Experts still aren't sure exactly how all *E. coli* outbreaks occur. It is not always known how certain food products become contaminated, or how soils and water carry *E. coli*. Even today there is much to be learned about how *E. coli* is spread.

E. coli is mostly spread through the food chain. That means you can get *E. coli* by eating infected food or drinking contaminated water. The World Health Organization, which works to stop the spread of dangerous bacteria like *E. coli*, makes the following recommendations when it comes to the safe handling and consumption of food.

- Make sure your food, particularly foods made with ground beef (e.g., hamburger) are properly cooked and are still hot when served.
- Avoid raw milk and products made from raw milk. Drink only pasteurized or boiled milk.
- Wash hands thoroughly and frequently using soap, in particular after having been to the toilet or after contact with farm animals.

- Wash fruits and vegetables carefully, particularly if they are eaten raw. If possible, vegetables and fruits should be peeled.
- When the safety of drinking water is doubtful, boil it; or if this is not possible, disinfect it with a reliable, slow-release disinfectant. These are usually available at pharmacies.

Diagnosis and Treatment

Most victims of *E. coli* have no idea what's wrong until they go to their doctors for help. And even then, it can be difficult for a doctor who has never seen the virus before to recognize it.

The way an *E. coli* 0157:H7 infection is diagnosed is through analysis of the patient's stool. Most tests for *E. coli* are ordered when a patient arrives at a doctor's office complaining of bloody stools and diarrhea and standard tests fail to pinpoint any other diseases. If a doctor suspects that *E. coli* might be the culprit in the patient's illness, he or she will request that a stool sample be taken. That sample is then brought to a laboratory to be studied. If the tests come back with positive results—that is, they detect *E. coli* in the stool—the patient is then diagnosed with an *E. coli* infection. Unfortunately, most labs don't routinely test stools for *E. coli*, so very often a person with *E. coli* will be told that the reason for his or her illness is unknown.

People who do come down with *E. coli* usually recover within five to ten days without any specific treatment. People with known cases of *E. coli* are usually given antibiotics to help them recover from the illness. Patients are told to avoid using antidiarrheal drugs—medications that prevent them from having diarrhea. This allows the bacteria to escape the intestines and blood instead of being trapped inside the body. While most people recover just fine, the elderly and the very young are at higher risk for developing life-threatening conditions. This is because these patients' immune systems—their bodies' natural infection fighters—are often not as strong. Weak immune systems are often unable to destroy dangerous invaders on their own.

Because the diarrhea is usually allowed to continue until the illness has passed, *E. coli* victims must be especially careful to avoid dehydration. Dehydration occurs when a person loses important body fluids. These fluids, especially water, contain salts and sugars that are needed for the body to function.

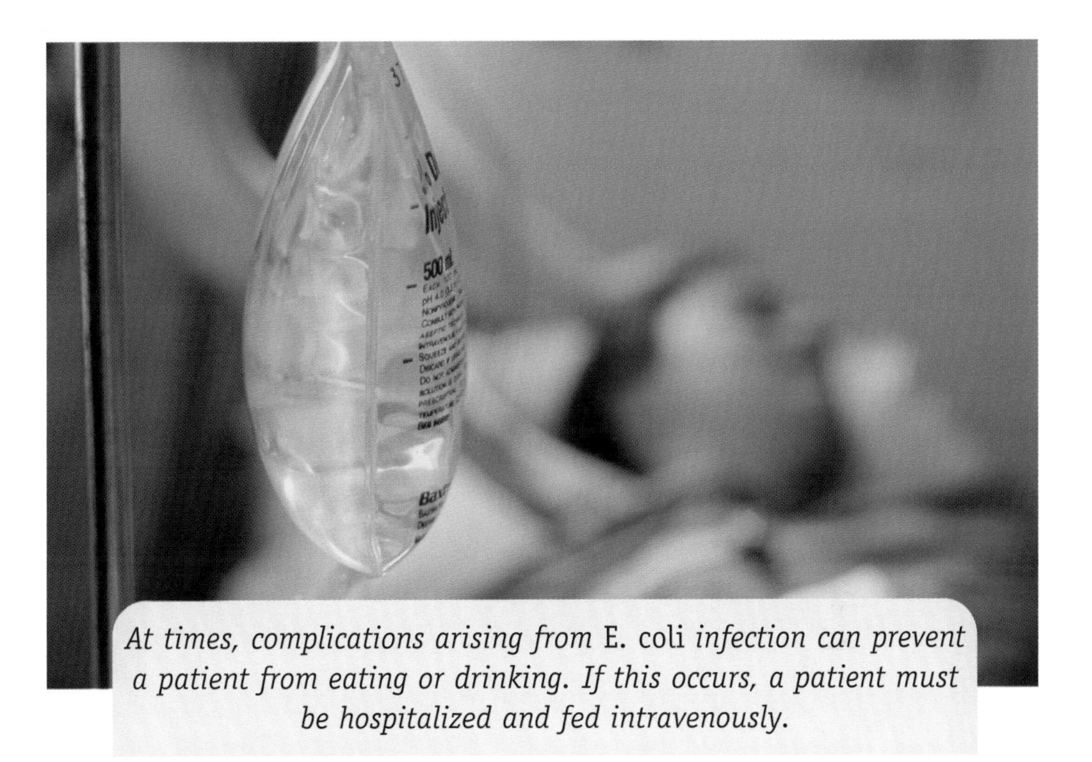

At times, complications arising from E. coli *infection can prevent a patient from eating or drinking. If this occurs, a patient must be hospitalized and fed intravenously.*

Diarrhea sends these fluids rushing out of the body. Severe dehydration is a potentially deadly condition.

The best way to avoid dehydration is by replacing the fluids and salts that are lost. For patients who have an appetite, this is fairly easy. Eating plenty of soup and drinking juices and sports drinks does the trick. But sometimes people sick from *E. coli* are not able to eat and drink on their own. When this happens, they must be hospitalized and given fluids intravenously. Intravenous fluids are delivered directly into the bloodstream through tubes.

5

THE FUTURE OF *E. COLI*

No one can say for certain what *E. coli* has in store for the future. There's no telling if new strains of the bacteria will emerge or if old strains will somehow mutate to become dangerous. Public health agencies have their hands full anytime they try to learn everything there is to know about a disease and how they can stop it. With *E. coli*, the story is the same. There's always something else that can be discovered and something new to be learned.

The Battle Has Just Begun

One of the main focuses for those involved in the fight against *E. coli* will be to find ways to decrease

E. COLI IN THE BODY

E. coli is a common microorganism in the human body. It makes up a tenth of a percent of the total amount of bacteria naturally found in the intestines. Surprisingly, its presence in the intestines is helpful, making it possible for people to absorb important vitamins and aiding in digestion.

the amount of meat that is contaminated by the bacteria. As experts at the CDC see it, this is very important because it's the best way to stop *E. coli* at its source: "*E. coli* 0157:H7 will continue to be an important health concern as long as it contaminates meat. Preventative measures may reduce the number of cattle that carry it and the contamination of meat during slaughter and grinding. Research into such prevention measures is just beginning."

Toward this end, experts are developing new farming and slaughtering techniques designed to make it harder for *E. coli* to enter meat during processing. A technique known as irradiation is also being investigated as a way to make ground beef safer to eat. Irradiation involves exposing the meat to intense light that can kill dangerous organisms. Experts will also be focusing on such basic things as understanding more about how *E. coli* lives in food animals. If they can figure out exactly how the

A linear accelerator like this one may one day help to irradiate food.

organism survives and the conditions it requires to flourish and eventually spread into the human food chain, then they can develop ways to fight it. The more they learn about the enemy, the easier it will be to destroy it.

Because raw food is another possible avenue for contracting *E. coli*, experts will also be working to find new ways to stop the bacteria before it can contaminate things like vegetables and fruit. This will be difficult, as the ways by which *E. coli* spreads to raw food are not yet fully understood. Stopping the spread of *E. coli* through produce will also be difficult because few people are aware it can be present on food that comes straight from the earth.

Educating the Public

Most experts agree that public education about *E. coli* is the key to prevention. For instance, public campaigns have and will continue to be organized to let people know how important it is to cook their food thoroughly before eating it. Many people have no idea that *E. coli* may lurk within the pink meat of their hamburgers. Through public health advertisements in magazines and newspapers, as well as on television and in local schools and libraries, experts can spread the word and potentially save lives.

New Technology

Scientists will also be trying to develop technology that will allow them to identify *E. coli* faster and easier. Researchers are now investigating ways to make *E. coli* bacteria isolation techniques more efficient. They're building special portable kits that can be used to detect *E. coli* in the field, directly in feces, in just a few hours, whereas in the past it has been necessary to bring samples all the way to the lab to be studied.

Doctors are also developing ways to make it easier to find *E. coli* that is actually in their patients. For now, blood must be drawn to determine whether *E. coli* is present, but less invasive techniques will make this

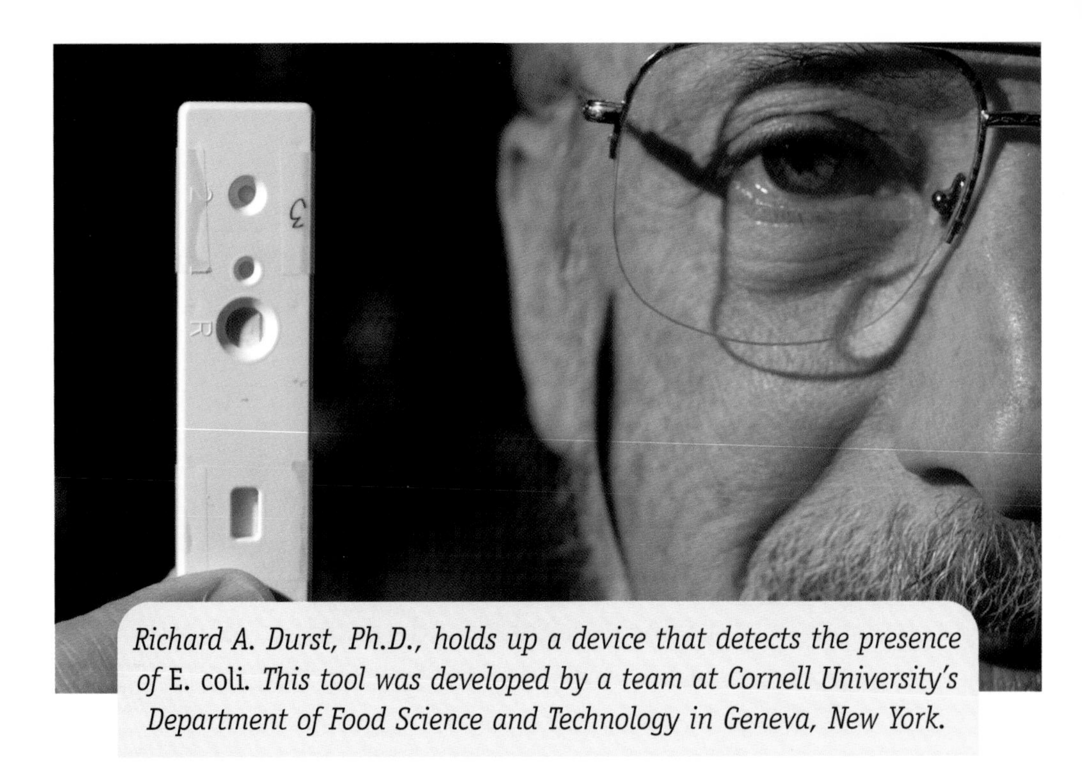

Richard A. Durst, Ph.D., holds up a device that detects the presence of E. coli. *This tool was developed by a team at Cornell University's Department of Food Science and Technology in Geneva, New York.*

unnecessary. One such way is through testing a patient's saliva for signs of the bacteria.

Better Communication

Public health organizations from around the world are continuing to work together in an effort to create an efficient international network that can identify and communicate about outbreaks. This will do wonders in terms of stopping outbreaks before they get out of control. Early detection of an outbreak can mean the difference between just one or two people coming down with an *E. coli* infection and hundreds or thousands of people falling ill.

In the Meantime

For now, of course, there's not much the general public can do about *E. coli* aside from using common sense when cooking and eating food—doing things like cleaning hands, cooking meat thoroughly, and washing vegetables. We all have to eat to survive. And we all have to drink. Humans come into contact with each other all the time. We can't change those things, nor should we. The fact is, people are going to contract *E. coli* once in a while, and as long as they do, some will become very sick and may even die. The best people can do is to use common sense to reduce the risk.

GLOSSARY

bacteria Microorganisms that are sometimes capable of causing harm to the human body but are often beneficial.

Centers for Disease Control and Prevention (CDC) An agency of the United States government that works to improve health and quality of life by preventing and controlling diseases, injuries, and disabilities.

contaminate To make impure or unclean.

dialysis A medical process used to clean waste products from the blood of people with non-functional or poorly functioning kidneys.

disease A change in the normal structure or function of any part of the body, characterized by specific unpleasant symptoms or signs.

epidemiology A branch of medical science that examines and tracks how diseases occur in populations.

hemolytic uremic syndrome (HUS) A rare kidney disorder in which cells are destroyed and the kidneys can stop working.

immune system The bodily system that protects the body from foreign substances.

infection The invasion of a host (a human being, for example) by a microorganism. Infections may eventually lead to disease.

intestine A body organ used to hold and excrete waste products.

laboratory A place used by scientists to conduct experiments and run tests.

microorganism An organism that can only be seen with the aid of a microscope.

organism An individual living plant, animal, or microorganism.

outbreak A sudden rise in the occurrence of a disease in a particular area or population.

pathogen Something that causes disease, like certain bacteria.

sanitary Clean.

sanitation The act of making sanitary, or clean.

stool sample A small collection of a person's feces used for laboratory analysis.

symptom Something that indicates the presence of a problem in the body.

World Health Organization (WHO) An agency of the United Nations whose purpose it is to promote physical, mental, and social well-being in people around the world. The organization works to strengthen nations' health services and prevent and control diseases, among other things.

FOR MORE INFORMATION

In the United States

American Medical Association (AMA)
515 North State Street
Chicago, IL 60610
(312) 464-5000
Web site: http://www.ama-assn.org

Centers for Disease Control and Prevention (CDC)
1600 Clifton Road
Atlanta, GA 30333
(800) 311-3435
Web site: http://www.cdc.gov

Environmental Protection Agency (EPA)
Ariel Rios Building
1200 Pennsylvania Avenue NW
Mail Code 3213A
Washington, DC 20460-0003
(202) 260-2090
Web site: http://www.epa.gov

National Institute of Allergy and Infectious Diseases
NIAID Office of Communications and Public Liaison
31 Center Drive, Room 7A-50
Bethesda, MD 20892-2520
(301) 496-1884
Web site: http://www.niaid.nih.gov

National Science Foundation
4201 Wilson Boulevard
Arlington, VA 22230
(703) 292-5111
Web site: http://www.nsf.gov

World Health Organization (WHO)
Regional Office for the Americas
Pan American Health Organization
525 23rd Street NW
Washington, DC 20037
(202) 974-3000
Web site: http://www.who.int/en

In Canada

Population and Public Health Branch
Bureau of Infectious Diseases
Health Protection Branch
Health Canada
Postal Locator: 0603E1
Tunney's Pasture
Ottawa, ON K1A 0L2
Web site: http://www.hc-sc.gc.ca

Web Sites

Due to the changing nature of Internet links, the Rosen Publishing Group, Inc., has developed an online list of Web sites related to the subject of this book. This site is updated regularly. Please use this link to access the list:

http://www.rosenlinks.com/epid/ecol

FOR FURTHER READING

Marsh, Carole. *Hot Zones: Disease, Epidemics, Viruses and Bacteria*. Peach Tree City, GA: Gallopade International, 1999.

Parry, Sharon, and Stephen Palmer. *E-Coli: Environmental Health Issues of Vtec 0157*. London: Routledge, 2002.

Ward, Brian R. *Epidemic*. New York: DK Publishing, 2000.

Yount, Lisa. *Epidemics*. San Diego: Lucent Books, 2000.

INDEX

CREDITS

About the Author

Chris Hayhurst is a freelance health and science writer living in Colorado.

Photo Credits

Cover, chapter title interior photos © Manfred Kage/Peter Arnold; p. 4 © Dennis Grundman/AP/Wide World Photos; p. 15 © Morocco Flowers/Index Stock Imagery, Inc.; p. 16 © Michael S. Yamasita/Corbis; p. 18 © Inga Spence/Index Stock Imagery, Inc.; p. 27 Environmental Protection Agency; p. 29 © Japack Company/Corbis; p. 33 © Bob Daemmrich/The Image Works; p. 36 © Custom Medical Stock Photo; p. 42 © Eric O'Connell/Taxi/Getty Images; p. 43 © Paul Poplis/ FoodPix/Getty Images; p. 44 © J. Reid/Custom Medical Stock Photo; p. 47 © IT Stock International/Index Stock; p. 50 © Charlie Neibergall/AP/Wide World Photos; p. 52 © Kevin Rivoli/AP/Wide World Photos.

Designer: Evelyn Horovicz; Editor: Eliza Berkowitz